Make Smoke, Burn Smoke

Biomass Gasification Primer

Smoke on Fire

By Doug Brethower

First Edition, July 2013

Doug Brethower
Rt. 1, Box 23F
Preston, MO 65732
freedombiomass@gmail.com
http://resiliencemovement.com
Technical editing and cover by JW Petermann

This work is licensed under a Creative Commons Attribution-NonCommercial-NoDerivs 3.0 Unported License.

You are free to share, copy, distribute and transmit this work under the following conditions:

Attribution — You must attribute the work by name "Make Smoke, Burn Smoke" and include the authors name

Noncommercial — You may not use this work for commercial purposes.

No Derivative Works — You may not alter, transform, or build upon this work.

Warning, burning smoke is an outdoor activity!
Experiment only when proper ventilation is assured

Incomplete combustion in a "make smoke, burn smoke" device is usually readily apparent by the thick smoke.

The thick smoke can be choking and poisonous to the point of being deadly in confined spaces. But that hazard is usually readily apparent.

A much more insidious hazard arises from incomplete combustion of charcoal. When a smoke burner design goes into char burning mode, the visible emissions remain clean, but the product of incomplete charcoal combustion can be deadly indoors.

Burning charcoal inside can kill you. It gives off carbon monoxide which has no odor.

Never burn charcoal inside homes, vehicles, or tents.

Make Smoke, Burn Smoke
- Biomass Gasification Primer

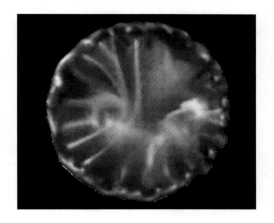

Smoke on Fire

by **Doug Brethower**

Acknowledgments

Special thanks to the many who helped raise awareness along the way:

Dr. Tom Reed
Dr. Paul Anderson
Raymond Rissler
Wayne Keith
Paul Wever
Mike LaRosa
Ed Torrey
Jeff Schneider
Joel Sommerville
David Yarrow
Lori Paro
Jim Hart
and

Mark, Franklin, Jason, Lloyd, Hugh, Dale, David, the Pease's and many, many others.

Special thanks to Joyce Riley of **The Power Hour** for having the courage to air what must have seemed crazy concepts on first hearing, to a worldwide audience. Their collective wisdom, insights, questions, and words of encouragement were foundational in the creation of this work.

And especially thank you for reading. May you enjoy much success "burning smoke".

Forward
By Jerry W. Petermann

There are few times in this modern world when one has a chance to be present when a revolution is at hand. Recently, I had the wonderful opportunity to be invited as a guest on the radio show The Power Hour with Joyce Riley. The co-host for that day was Doug Brethower.

I was in awe. Here was the fellow she had recently interviewed, a personal friend, who was *the* local authority on wood gasification and something called Bio Char. Since I had heard one of their interviews on the subject, I couldn't wait to call Doug after the show and get to the bottom of this "New" natural resource. There was a project I was working on and this was THE answer to a huge problem.

Although Doug freely offered his technical know-how on my venture, I quickly discovered his overall vision: teach the next generation and those of us still responsive to a really great revolutionary Bio Green Solution!

This Primer is his heart and soul. This work crystallizes his philosophy on Bio Char and "free" energy from bio mass so the guy next door or the girl across the street can grasp the concept. This work instructs, lifts up and charts the course ahead for a grassroots revolution!

Make Smoke, Burn Smoke is not an encyclopedia! It's to the point, an easy read and just about everything you need to know to join in! Though it is a "Primer", it may well become the text reference on the subject.

So, kick back, put up your feet and soak it in. Then, get up and make it happen! It's all in there....so get busy and:
>   Make Smoke, Burn Smoke!

Table of Contents

Chapter 1 - Why burn smoke?
-- Battery Storage for Solar
-- Clean Energy
-- Fire as a Creative Act
-- Economies of Scale
-- We Won't Run Out of Trees
-- Waste Biomass
-- Energy and the Balance of Nature

Chapter 2 - Terra Preta Authentica, Clay Smoke Burners
-- Thank You Lori!
-- Basic Design Details
-- Fuel Cell
-- Usage
-- Design Notes
-- Guidance
-- Fuels make a difference
-- Charcoal Makers Guild

Chapter 3 - Tincanium Smoke Burners
-- History and benefits
-- Simplest Possible Smoke Burner
-- Self-Extinguishing Smoke Burner
-- Purchased Parts Smoke Burner
-- KeySTove LX Advanced Design
-- 30x55 Barrel Sized Smoke Burner
-- Continuous Biochar Production with heat
-- More about Fuels

Chapter 4 - Engineered Designs
-- ARTI Sugarcane Leaf Retort
-- An•Thro•Soil•1 "Grassifier"
-- Charring vs Burning
-- More about Charcoal
-- Fan Powered Smoke Burner

Chapter 5 - Burning Smoke in Internal Combustion Engines

-- The Promise and the History
  -- MGS-80 Sawdust Gasifier
  -- Stationary Systems
  -- The Wayne Keith Gasifier for Modern Engines
  -- Mike LaRosa and LaRosifier

Afterword

Addendum
  -- Unused Ag Wastes
  -- Stove Emissions Standards and Testing
  -- Biochar Durability in Soil
  -- FEMA Gasifier
  -- CLEW Large Wood Waste Processor
  -- Power Grid Vulnerability to Solar Flares
  -- Global Atmospheric CO2 Readings

Recommended Readings
  -- Understanding Stoves - Bhaskar Reddy
  -- Micro-Gasification: Cooking with gas from biomass, Christa Roth
  -- On the WWW, Aprovecho.org

The Final Word
by David Yarrow

The last page of the book
109

Introduction
by Doug Brethower

This book focuses on pictures, diagrams, real world examples and general rules of thumb.

"Woodgas is simple, once you understand it" - Wayne Keith. But there remains much to be learned. The learning will come through sharing and individual experimentation. Like the phone and the pc, this is a technology poised to leap forward by the "network effect". The more people using it, the more useful and valuable it becomes.

My intent in writing this book is to share my experiences and revelations, to help others see the world I envision through the smoke colored lens.

Formula's, technical jargon, and engineering calculations, are purposefully omitted unless relevant to exactly the issue discussed.

This book is rife with well-researched opinions expressed by the author not referenced to an exact source. I hope you enjoy it in the spirit of learning through someone else's real world experiences.

Chapter 1

Why would anyone want to do this?

Biomass is "Battery Storage" for Solar Energy

The sun delivers humanly inconceivable amounts of energy to our planet daily.

A portion of that energy is captured and stored in plants and trees in nature's finest hydrogen-carbon fuel cells. Every growing plant is a "power plant".

Man has yet to match the simple perfection of wood energy storage with any other form of battery or fuel. Kept dry, it stores indefinitely. When needed, it may be catalyzed for use in accordance with man's wisdom. Wood gasification done correctly is free energy, free as in beer, and free as in freedom to live comfortably within our means.

Clean Energy

Burning the smoke extracts more energy from the resource and cleans up the dirtiest part of the process of burning wood, the smoke.

To make smoke and burn smoke, the gasification process controls heat and oxygen levels at a point that essentially "boils" the hydrogen and lighter carbon molecules out of the cells of the wood in the form of a rich smoke.

Burning the smoke with fresh oxygen reconstitutes the molecules in a flame, forming carbon dioxide and water, both good for plants. This is the cleanest way yet devised to produce useful energy from a wood burning appliance. With precise control of fuel and air, conversion efficiency approaches that of the best coal-fired power plants.

No oil spills, no radioactive wastes, no massive transportation and distribution infrastructure is required.

See the emissions chart in the addendum for lab hood emissions testing of "make smoke, burn smoke" appliances relative to other clean cook stove technology. **Spoiler:** smoke burners, particularly when saving the char, have by far the cleanest emissions of any biomass cooking appliances.

Fire as a Creative Act

The amount of energy in wood, when converted efficiently, must be seen to be believed.

A good double handful of dry twigs can boil a quart of water for thirty minutes and leave behind useful charcoal! Burning that same double handful of dry twigs out in the open quickly leaves a small pile of ash with little else to show for the effort.

Boiling the most volatile elements from wood for use as flare gas, while saving the hard black carbon structures, not only cleans up emissions, it also completely changes the modern paradigm of fire from a destructive, to a creative act.

The hard carbon that remains, charcoal, has many uses:
- Medicine - Hippocrates recommended about 400 BC, poison control and digestive aid
- Liquids filtration - Began in the BC era in Rome. Salvaging marginal wine reputed to have saved a few necks. Today

many uses including under-sink water filters
- Soils improvement - Fertile Terra Preta Soils date back 6,000 years. Modern re-discovery of soil charcoal is termed "biochar"
- Component of black powder, 7th century China to today.

Variable internal pore structures and a fixed electric charge on surfaces make charcoal the most unique adsorbent known to man.

To the extent the smoke from charcoal production is flared for use in cooking, drying or warming, the conversion of biomass to charcoal is not merely zero waste, it is a net gain. When the charcoal is used for soil improvement (biochar), the process is carbon negative and self-reinforcing through greater plant growth.

Growing more biomass per acre means more useful energy captured from the sun. We waste this valuable free energy and subsidize non-renewable sources at our peril.

Economies of Scale

Energy from "making smoke, burning smoke" is "thin" technology. It scales in usefulness down to the level of the individual. As complex systems run into complex problems, low tech becomes the latest high tech. Burning smoke requires some understanding, but like riding a bicycle, it becomes effortless with practice. Designs can range from utilitarian to steam punk to Rube Goldberg, from individual cooking requirements to community power generation.

Vehicle gasification systems built by individuals convert wood to mileage at the approximate rate of one pound of wood per mile traveled for light vehicles. A large feed sack of wood chunks is roughly to equivalent to 2.5 gallons of gasoline.

We Won't Run Out of Trees

Should the day ever come when an excess of woody biomass is not a problem to be chopped up, burned, landfilled, or left to rot, there are many options to continue down the path. About 4 cords of wood, a couple of big trees, have the energy to power the average vehicle the average miles per year.

Fast growing stemmy biomass, bamboos, canes, and warm season grasses, can provide up to 10 tons per acre per year **from lands too poor for cropping**. That is about 20,000 vehicle miles per acre EVERY YEAR!

Well planned human interaction, plantings, clearings and thinnings, could easily double the total amount of solar energy captured by biomass per unit of land virtually everywhere in the US. Converting solar energy directly into food, feed and fuels beyond what is needed locally, is infinitely sustainable wealth expansion.

Waste Biomass

Biomass waste is a nebulous term. None of it really goes to waste, nature will consume all of it one way or another. **Wildfires are the most destructive example.**

Planning enough biomass for human energy use is a more

accurate description for the future.

Man's Energy Needs Have a Place in the Balance of Nature.

That place is not necessarily shackled to a gas pump.

Early visitors to the New World reported gleaming cities of great wealth in the Amazon basin. More recently soil scientists studying the mystery of fertile soils being mined as potting soil in the middle of the jungle, discovered high concentrations of char, clay shards and bone fragments.

Living in relative ease, perhaps the ancient residents who created Terra Preta soils had discovered a near perfect balance of nature feeding man, man feeding nature. Terra Preta is Portuguese for "Pretty Earth".

Chapter 2
Terra Preta Authentica

How did they make charcoal 6,000 years ago in Amazonia?

We can only guess from the materials, tools and needs of the day.

Cooking with clean, efficient clay vessels could have provided the easy life style noted by early visitors who told of "gleaming cities", not "dim smoky encampments". Convective cooling of dwellings may have been powered by natural draft feeding clay charcoal kilns. The legend of El Dorado, may not have been a legend.

Based on the reports of "gleaming cities of great wealth", we will assume this civilization had discovered how to do energy right, in the cleanest, least wasteful, most useful way possible.

They were known to fashion clay. Clay is an excellent construction material for simple devices that "make smoke, burn smoke"!

Clay is:

- inexpensive
- easy to form into custom shapes
- handles high temperatures well
- doesn't rust or wear out with repeated use like tin cans (covered next)
- exudes radiant heat long after snuffing to heat a space while conditioning the charcoal.

Thank You, Lori!

Clay "smoke burner" parts before firing

Very special thanks to artist-friend, Lori Paro for taking hold of the concepts, making the first real working example, and particularly for sharing the knowledge, lore, and enthusiasm with her art students. The kids dubbed their devices "flux capacitors" in honor of the movie "Back to the Future".

Lori shows the artistic way to hold a torch

Basic Design Details

The basic design is three pieces, a base, fuel cell, and chimney. The base suspends the fuel cell allowing free flow of air upward from the bottom of the fuel cell, to the smoke front, to just below the chimney where the smoke is flared, then on up the chimney.

With a big enough plate around the chimney, a plancha cooking surface is created. In a larger design, the chimney need not necessarily be centered (Hint).

The trick is getting just enough oxygen to a downward progressing front to boil hot thick smoke out of the biomass material. The hot smoke rises, then ignites when a second round of oxygen is introduced at the top of the fuel cell below the chimney. The flame exiting the chimney induces draft, creating continuous upward air movement.

The pile between the downward progressing smoke front, and the upper air holes is oxygen starved. So the "boiled" biomass, the charcoal, is saved. After the process runs to completion, all the way to the bottom, the charcoal will begin to burn from the bottom up if the holes are not capped or the charcoal dumped.

**Note that although the processing of charcoal from bottom up appears to be running cleanly, it is likely that carbon monoxide is being produced at elevated levels. Carbon monoxide is odorless and potentially deadly indoors.

Fuel Cell

The base and the chimney are fairly intuitive. Air hole(s) in the support base should be below the bottom of the fuel cell and at least half the face area of the tightest chimney restriction. This seems to happen naturally among the small sampling of artists who built them. The fuel cell is the most error prone component. It is shown below upside down to display a successful pattern of holes in the bottom. Note the many small holes versus a few big ones. Also that they are distributed evenly across the entire bottom surface. At least "kind of" evenly - this is art!

The more even the air distribution through the pile, the more cleanly the pile processes. The base supports the holes up off the ground allowing air to enter. Tight spacing between the fuel cell and the support base pre-heats the second round of air before it is injected to burn the smoke. Pre-heating allows higher rates of air injection with less worry of quenching the flame - "blowing out the candle".

Inverted fuel cell. It is flipped over in actual use.

USAGE

To fire up a batch:
- Fuel is added, possibly tamped or fluffed depending upon fuel density and past experience --**NOTE:** a fuel level that blocks the top holes causes a smokey mess
- The cell is lowered into the base
- The entire top (and only the top) of the biomass pile is lit as completely, evenly, and quickly as possible
- As a nice flame begins dancing across the top of the pile, the chimney is lowered into place
- The process continues unattended from this point, akin to burning a field of dry grass into the wind

A really great feature of these appliances is a nearly constant heat output during runtime without tending, much like a burner on a stove

Runtime processing

*Visible smoke at any time indicates a problem. Done correctly these burn clean!

- There will be a noticeable flame when looking down the chimney
- The flame will rise as the process gets rolling, stabilize, then begin descending as the cell exhausts its fuel supply
- Experience determines the best time to finish the process by dumping or capping

Lori's Improved Biochar Maker

Design notes

Lori uses oven mitts to remove the chimney and dump the glowing embers and char into a sealable metal container. Then she reloads and re-lights. With a little design effort and plugs for both ends, the entire process could be capped instead of dumped, giving durable radiant heat over time. Two or three systems running in parallel would allow some cool down time and take a lot of the peril out of the dumping operation.

The small design pictured in the lead-in to the chapter heats about like a candle. It runs for twenty to thirty minutes on a single charge of crushed dry corn husks or almost 45 minutes on a small handful of finely crushed corn cobs. The one pictured above warms Lori's workspace in the garage nicely on a cool day. Her flowers love the biochar!

Guidance and Common Miscues

Upper holes approximately twice the total face area of the bottom air feed holes seems to be a good ratio. Either twice as many holes, or the same number of holes but twice as big. Excess upper air, as shown in the inverted fuel cell picture, is better than not enough. Not enough and smoke is created faster than it can be burned.

More air feeding the bottom of the pile makes startups easier and the process completes faster with higher heat output. Less air feeding the bottom makes starting more difficult and gives longer run times at lower heat output.

Accelerant soaked material as the final layer on the top of the pile in the cell is a great aid to fast clean starts. A waste vegetable oil-alcohol mix works well. Small pieces of high grade tissue paper, or some crushed stemmy grass incorporated into the starting layer helps the flame spread evenly and quickly.

Pouring accelerant directly on the pile is a common misstep. The starting flame may race down into the pile chasing the accelerant, then burn outward from there, caving in and creating a smoky mess.

Fuel makes a difference!

- Type
- Size
- Shape
- Packing/fluffing
- Moisture content

Changing any one of these can change a working design into a non-working design.

Tight spacing between the support base and fuel cell pre-heats the second round of air that burns the smoke, reducing the risk of quenching.

Taller chimneys provide more draft, up to a point. A height twenty times the chimney diameter is recognized as the point of decreasing return.

Lori's designs generally handle many fuel types, shapes and sizes well. They convert hardwood pellets or crushed stemmy grasses, into wonderful radiant heat and biochar.

Clay, yay!

Notes on Charcoal Making

Ancient charcoal makers, known as colliers, held "guild" status amongst peers. Up-converting wood was a combination of art and science, tuned by years of practical experience.

Two broad categories of charcoal are oxic and anoxic. Anoxic means in the absence of oxygen. Anoxic processes retain more weight by virtue of retaining more tars in the product, making a better fuel. Oxic charcoal is processed in the presence of oxygen, like in a smoke burner. Oxic produces a cleaner, lighter charcoal product.

Chapter 3
Tincanium/Refugee Grade Smoke Burners

Biomass Energy Foundation Combined Heat and Biochar Camp, February 2011

One reason smoke burner appliances are like snow flakes, no two alike, "playing with fire is fun!" Another reason is that this is an evolving art.

Modern day smoke burning pioneer Paal Wendelbo is notorious for his hours spent peering into the flame, noting colors and shapes that indicate how well a specific fuel is operating in a specific design. Paal learned the basic principles during WWII German occupation of Norway. Every bit as essential as the actual heating, cooking, and boiling of water for survival, was concealing the activity.

Paal Wendelbo and Dr. Paul Anderson (second from right in picture above) are pyroneers in moving cooking scale "tincanium" smoke burner technology forward in the present day among the half of the world's population that remains free from petroleum rule.

Compared to rocket stove appliances:

- NO FIRE TENDING
- Consistent output, more like a standard burner
- Done correctly - MUCH cleaner emissions
- Higher fuel efficiency with less skilled operator
- Saves the charcoal!

"Woodgas for cooking - Third World Cooking", from Dr. Paul Anderson, aka "DrTLUD" drtlud.com

The qualities researched by third world cooks in a cook stove are, by order of importance:
1. fast cooking
2. easiness and comfort of use (untended cooking, burner like)
3. durability
4. wood savings
5. smokes and noxious gas emissions reduction
6. stove looks

Although biomass cooking should never be done indoors, it is done indoors every day in many third world countries. When indoors is the only choice, smoke burners are the best choice. They offer the least tending during runtime, most efficient use of combustible biomass, and lowest emissions. Smoke from a cigarette is more noticeable in a room than emissions from a high performance smoke burner.

Building and testing with available scrap cans (obtainium), improves the art and science while spreading basic understanding of the processes. The basic tenet of tincanium is that given whatever collection of cans, and common hand tools, make something that converts a readily available fuel efficiently and cleanly into heat and char. Small battery operable fans (salvaged from old desktop computers) are allowed in friendly learning competitions sponsored by the **Biomass Energy Foundation**.

The key search term for finding more basic information is TLUD, Top Lit Up Draft.

Simplest Possible Tincanium Smoke Burner

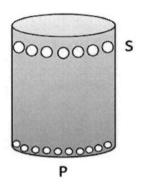 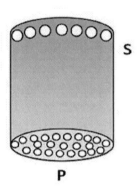

The primary air holes from the sides are suitable for small size TLUDs.   *The primary air holes at the bottom are suitable for large size TLUDs.*

Source - *"Understanding Stoves"* Dr. N. Sai Bhaskar Reddy, 2012

A single can punched with holes makes a reasonably effective "smoke burner" useful on a calm day. Notice the similarity to the "fuel cell" in the clay designs presented in Chapter 1.

The obvious next step, just as in clay, is to fashion a support base or enclosure from a larger can, that either slips over this cell or this cell slips into. A two-can burner in other words, squawk. A simple improvement is finding a shorter wider can, cutting a hole centered in the bottom, and inverting it over the fuel cell to concentrate the flame.

Source: Micro-gasification: Cooking with gas from biomass,
Christa Roth
GIZ HERA January 2011

The two pictures on the previous page are taken from much

more complete stove study references that go into greater detail and are highly recommended for the serious student.

More ideas for exploring the joys of tincanium from a 1959 US Patent #2877759 drawing.

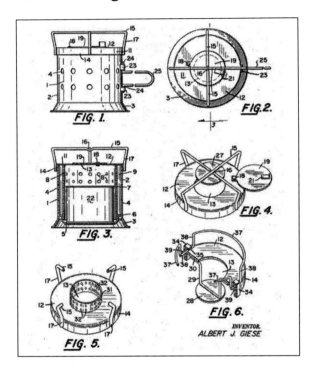

Self-Extinguishing Smoke Burner

After the smoke is boiled from the pile, the remaining charcoal begins burning from the bottom up. The air fuel ratio that was perfect for boiling wood into smoke is too low for clean combustion of charcoal. Incomplete combustion creates an excess of carbon monoxide while turning the valuable char into ash. Both bad things.

Although all fires eventually go out when the fuel is consumed, the challenge was to build a device that would "self-extinguish" after smoke burning, before consuming the char.

The scraps were a stainless drink mug missing the handle, a drop cut of 4" vent pipe, the bottom from a can that fit into the vent pipe, a coffee pot lid, an old electric heating element, and a short section of 1" copper pipe. Good luck finding copper in a scrap bin these days. But a short section of any 1" metal pipe should work about as well.

The stainless drink mug is double wall, vacuum sealed, essentially a cup within a cup. Punching some holes in the bottom of the outer, and the bottom of the inner, then a ring of holes on the inside near the lip makes a simple smoke burner that works reasonably well on a perfectly calm day when elevated slightly above the ground to allow air flow.

For operation on windy days, the base concept is extended along the lines of the clay smoke burners. Enclose the top to concentrate the gas flare, add a chimney to induce draft.

Taking these simple idea a few steps further, creates a smoke burner that reliably self extinguishes, even on a windy day, leaving almost all the char intact.

The few steps further are:
- a shroud to help retain heat and entrain air flow
- another set of holes above the inner cup bottom holes
- a directional base to prevent wind pushing oxygenated air into system after the flare dies down

Self-Extinguishing Smoke Burner in Operation

Conceptually what happens is the chimney cools as the flare dies down. The more it cools, the more it acts as a cap instead of a chimney. Warm air still rising inside the shroud, seeks the first path of least resistance, the upper level holes which feed the inner top ring of holes that once fed oxygen to the flare. Now they quench instead of flaring. The system quenches just progressively enough to finish the char without consuming it.

There is a little more to it, the chimney is partially shrouded as shown in the pictures. The proof is in the operation, and operationally, it works! This design very reliably self-extinguishes when processing hard fuels like wood and crushed corn cobs.

The processor and shroud in two parts. Not clear from the picture above is a ring of holes just below the top rim of the mug (on the inside only) for fresh air to burn the smoke.

Shroud details, the chimney rests on the raised hole, the cap nests into the top locking the chimney in place

Notice how bottom holes are tight to center. The upper row of holes on the side of the mug only penetrate the outer skin.

Purchased Parts Smoke Burner

"With the air of freedom the flame grows bright" - Queensryche, Take Hold of the Flame

Purpose is spreading tincanium skills to those not experienced in the black art of dumpster diving.

Objective is a consumer friendly design collected from shelves of a local hardware store to quickly and easily create an effective smoke burner.

This is a slim variation on the clay designs, made possible by a food storage canister that slips perfectly into the throat of a gallon paint can and remains slightly elevated above the bottom of the can.

Parts List:
- One gallon paint can
- Stainless food storage canister to fit inside the paint can opening (one gallon size?)
- Metal toilet brush holder
- Roof vent collar (if memory serves)
- Furnace cement

Fabrication:

- The food canister - punch or drill a base hole pattern in the bottom of the food canister, and a hole pattern at least double the total face area of the bottom holes near the top.
- The paint can - drill or punch enough holes to feed all the holes in the food canister around the bottom of the can
- Toilet brush holder - knock out about a 2" hole centered in the base to feed this "chimney".
- Roof Collar - slight inward bending so it fits tightly to the top of the paint can.
- Furnace cement, use like putty to seal and stick the brush holder over the opening in the roof collar.
- During the first run, the furnace cement bakes the chimney to the collar.

Fabricated Parts Smoke Burner

Chimney Assembly

Bottom View

Open Top View   Running

Note that the opening in the bottom of the paint can IS NOT REQUIRED for this design to operate

Note also that the airflow into the bottom of the fuel cell is fairly restricted. This gives long run times at low power. Higher power output (more holes) are possible but may overheat the bottom of the chimney. Depends upon the quality of the metal.

KeySTove vortex swirl looking down the chimney

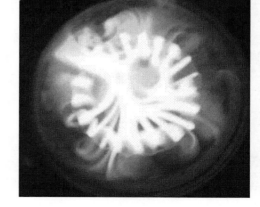

KeySTove LX - Purchased Parts Tinkering Toward Improved Designs

The name is a jab at the highly publicized and controversial Keystone XL project. This smoke burner operates at the opposite end of the scale from the cross-border pipeline, and has a much higher overall efficiency and practicality.

The US annually wastes and landfills more biomass btu than the net energy output expected from the massive carbon extraction project. Instead of rallying **against** tar sands, **promote** responsible small scale use of biomass energy. We need energy, it must come from somewhere.

The KeySTove LX was a featured project on instructables.com for the month of December 2011. It has been visited almost 20,000 times as of June 2013.
  -- http://www.instructables.com/id/Durable-Biochar-Producing-TLUD-Camp-Stove/

Parts list:

- 2 liter stainless steel vacuum insulated coffee serving pot
- Paint can
- 5" length of 4" diameter aluminized vehicle exhaust pipe
- 1" thick ceramic wool insulation (found at insulation distributor warehouse)

Fabrication:

- Wreck the double wall serving pot down to the bare vessel, removing the false bottom

- Punch holes in the outside bottom to suit, feed about twice the air up the sides as to the bottom
- Punch holes in the inner liner bottom. Offsetting the holes from the outer bottom reduces risk of hot char falling out the bottom as processing completes
- Drill the upper holes all the way through both skins. Drill them about where the 4" pipe will sit and the chimney effectively seals the holes in the outer skin
- Crimp the top inward with pliers to concentrate the gas as much as possible

Slipping the exhaust pipe in place as a chimney at this point creates a nice smoke burner just by elevating the serving pot on three small rocks so the air can flow into the base. However the outer skin gets hot enough to lift fingerprints at the slightest touch. That is a little too dangerous to be everyday useful.

More fabrication:

- Cut a piece of the ceramic wool insulation to fit the inside of the paint can
- Cut about a 3 3/4 inch hole centered in the bottom of the paint can.
- Snip the edges of this hole slightly so the 4" pipe is held snug as it is forced through
- Force the original open end of the paint can down over the neck of the serving pot
- Force the chimney down through the hole until it rests firmly on the top rim of the serving pot

Fill the serving pot about half full for the first run.

Startup Procedure

The reducing neck of the serving pot creates an awesome vortex effect at runtime, but makes this one of the more difficult smoke burner designs to start.

Overfilling, a common error among the untrained in all smoke burners, is no less common for this design. It works fine filled half to three quarters full. An accelerant soaked top layer is highly recommended for fast, clean starts.

If starting with a propane torch, a quick hit with the torch works best. Maintaining the torch flame too long burns off the accelerant layer without allowing the upward airflow to begin. A single sheet of lightly wadded toilet paper lit and dropped in works about as well as a torch. When the flame on top starts, patience is a virtue versus poking embers around, based on the experience of thousands of starts.

A removable chimney, like the bowl brush holder chimney depicted earlier is a superior aid for fast starts. Remove it when the flame becomes a durable swirl for fast, virtually smokeless startup.

KeyStove LX

Micro-Gasification Camp Stove

State of the Art
Top-Lit-Up-Draft "TLUD" Design

- Clean burning, no smoke, little flame
- High efficiency, saves char, carbon negative
- Operates reliably on hardwood pellets, chips, wheat straw, corn stover, cobs, brushpile stompage, dried biomass fuels

KeySTove LX Parts

KeySTove LX Bottom View of Serving Pot

Water Boiling Setup and in Operation

A 16 ounce cup of wood pellets, about half a charge, brought a quart of 55 degree F water to boil in 11 minutes and boiled down to 10 ounces of water by the time the flame started dying down 35 minutes later.

Coarse home video at
http://www.youtube.com/watch?v=yj4WfWGk46g

Charcoal Burning in the KeySTove LX

Normally the KeyStove is snuffed by capping both ends when the flame dies into the char bed, to preserve and condition the char.

In an emergency, when heat may be more important than another batch of charcoal, a simple attachment can switch the KeySTove LX from a smoke burner into a decent low heat output char burner. This works great for a hand or foot warmer during outdoor activities.

A can or in this case the bottom section of a toilet brush holder can be fashioned into a restriction to reduce air flow to very slowly reduce the char to ash. The restriction is put in place when the flame descends into the char bed. On the first test it ran overnight and was still producing a pleasant low heat almost like a catalytic heater the next morning, with char remaining.

NEVER burn charcoal indoors!! It can kill you. See the warning label on a bag of charcoal for more details.

Turn a KeySTove LX smoke burner into a low power char burner by reducing airflow

A Smaller Simpler Variation of the KeySTove LX for backpack duty

Scaled Up 30x55 Smoke Burner

The first person widely reported making garden scale biochar onsite was a fellow in Florida using a 55 gallon burn barrel with a chimney and lighting on top. Converting a 55 gallon drum into a smoke burner can be that simple on a good day.

For making useful quantities of biochar onsite, 55 gallon drums are hard to beat in terms of price, availability and performance. The designs are about like snowflakes, no two alike.

In this example, the growers wanted a setup that could keep a person busy all day making char and they had plenty of barrels. The idea was to make a single outer shroud with chimney that can be moved from 30 gallon barrel to 30 gallon barrel, with each 30 gallon barrel being capped when it finishes.

Parts List:
- 2 each 55 gallon barrels
- At least one 30 gallon barrel
- More 55 gallon and 30 gallon barrels as desired or available
- 2 each 6x8 flue pipe adapters
- 3 each 24" 6 inch flue pipe sections
- 2 handles, 2 wires or springs, 2 eyebolts with nuts
- Self-tapping sheet metal screws

Fabrication:
30 Gallon Barrel Fuel Cell
- The top needs to be open, remove the lid or if a solid barrel, cut away most of the top
- The bottom needs some holes punched in it to suit the biomass feedstock. Generally more holes for more compacted fuels, fewer holes for more open fuels. Too many holes in a hot burning high quality fuel risks melting the shroud barrel (almost happened in this case)

55 Gallon Barrel for Capping
- Needs to have an open top for sliding over the 30 gallon barrel following the processing
-

55 Gallon Shroud Barrel
- Open top for sliding over the 30 gallon barrel
- Cut a 5" hole centered in the bottom of the barrel
- Attach some kind of handles, eyes or hooks to ease the process of removing the barrel following processing
-

6x8 flue pipe adapter
- Snip and fold the entire 8" mating section inward
-

One section of 6" flue pipe
- Attach handles to simplify removal of the hot chimney following processing
- Optional, attach hooks or eye bolts for holding anchoring wires/springs

Putting it together
With the 55 gallon shroud barrel opening facing down, center the snipped 6x8 flue pipe adapter over the hole in the bottom and attach it with sheet metal screws. Attach lifting eyes to the shroud barrel.

Put the three sections of flue pipe together with the anchoring section in the middle. Slide the assembled section into the connection of the second 6x8 adapter.

Operation
Find a smooth flat location, away from structures, trees, and flammables and fully open to the sky. Preferably the location

should be near dirt that can easily be scooped to help seal the bottom of the cap barrel. Find some fire resistant spacer material, bricks, blocks, short pieces of pipe that will position the top of the 30 gallon barrel within about an inch of the bottom of the inverted 55 gallon shroud barrel when operating. Keep in mind the shroud barrel will be elevated about an inch or two to allow process air flow. Some 1" thick boards work well for elevating the shroud barrel.

Fill the 30 gallon barrel with fuel, lightly sprinkle some accelerant on top, light it. When flame is looking good, cover the 30 gallon barrel with the shroud barrel and attach the chimney.

30 gallon fuel cell finished

6x8 adapter snipped and folded

Shroud Barrel Finished

Chimney Finished

Fuel cell filled

Lowering shroud barrel in place

Processing

Capping

Finished Dry Charcoal

Two man lift of hot shroud barrel

Discussion:

Different numbers of holes can be punched in the 30 gallon barrels to handle various fuels. In this case holes were punched for wood chips, and the feedstock instead was mostly hardwood flooring and bone dry brushpile stompage. The top of the shroud barrel glowed red hot and looked to be in danger of melting early in the process. Note the tan circle around the base of the chimney in the processing picture.

The first snuffing barrel was a little shorter than the shroud barrel, so some of the char burned to ash (the white in the middle of the char picture) while a longer barrel was sorted out of the pile.

Not all 55 gallon barrels are the same dimensions.

Other than the hot spot below the chimney, the rest of shroud barrel remains amazingly cool, hot to the touch, but losing no paint.

Note how the handles on the chimney and eye bolts on the shroud barrel double as anchor attachments. Also note the eye bolts allow sliding rods through for a simple and relatively safe two-person lift of the shroud barrel off the hot char barrel.

Continuous Process Biochar Production

A good hand with a shovel can make about seven gallons of biochar and a few hundred thousand btu per hour on a calm day or evening.

Praise to Joel Sommerville for this witty invention designed and built to aid his growing endeavors. Small chunks of wood are shoveled in at the top. Cooled biochar is augered out of the bottom. Joel fashioned it after making his own char in burn piles, paying attention to how char gets saved in nature.

Slipping a scrap pressure tank with an added concentrator ring over a thicker wall char holding tank with strategic air control makes it work. Just like in nature, charcoal is shielded from oxygen by the flame above. The added trick is removing the charcoal from below, while the flames burn above.

This one is shown just in pictures and notes rather than build instructions, because the concepts are far more important than the fabrication details. This design is awesome in it simplicity and usefulness.

Shroud

Char Holding Tank

Holes and tubes for boiling wood, four fins standoff the shroud to supply air over the top to burn smoke. Charcoal cools as it descends, heating the supply air.

Holes and Tubes for Boiling Wood   **Four Fins Standoff the Shroud**

The arrangement of holes and lengths of tubes assure even charring across the entire circumference. The fins stand the shroud off just far enough to allow a smooth flow of air for smoke burning. Char cools as it descends, transferring heat to the incoming air entrained between the shroud and the char holding vessel.

Auger Outlet   **Auger dumps into wheelbarrow**

For first startup, the char holding tank must be filled to the level of the ports with something that will auger out.

Video link http://ww.youtube.com/watch?v=HdRlEjI_AJE

Youtube channel **Joel C**

Fuels
Joel's processor works great on small chunks of wood that are easily shoveled.

It was mentioned earlier, bears repeating here, that the target fuel is the most critical consideration when fabricating smoke burners.

- Type - wood, stemmy grass, seeds, nuts, pits, dry dung
- Size - infinite variations, the less energy used for processing the better (energy in vs energy out)
- Shape - some shapes tend to shingle and pack blocking airflow
- Packing/fluffing - same idea as shape, air flow through fuel pile is critical
- Moisture content

Changing any one of these can significantly alter the performance of a natural draft smoke burner.

Notice also that in all of these designs, **no external energy is required** other than a match or torch to kickstart the action. With Joel's continuous processing, one start can make a lot of char.

Chapter 4
Engineered Designs

ARTI Sugarcane Leaf Retort

Appropriate Rural Technologies Institute, Maharashta, India Charcoal Retort

Creating value from otherwise useless waste is true wealth creation. Sugar cane leaves aren't good for animal feed and don't break down well in soil. They were simply piled up and burned to get them out of the way for the next crop. An estimated four million tons per year were going up in smoke in the Maharashtra state of India.

ARTI, Appropriate Rural Technologies Institute in Pune developed a small retort system powered entirely by the sugar cane leaves that converts them to charcoal. Charcoal is a traditional fuel there. The charcoal powder, with a little binder, is then pressed into large pellets for use in charcoal stoves.

A charcoal retort is a furnace that holds a container of material to be physically converted by heating. In this case cane leaves are placed underneath an arrangement of seven cans filled with cane leaves and lit.

A Clay Retort Furnace with cans loaded and lid off

A few small holes in the lower circumference of each can allows the volatile smoke to escape as the can heats. The exiting smoke flares into a flame, helping power the conversion. With a little experimenting the operator learns how much fuel is required below to completely char the cane leaves in the cans.

This simple bit of engineering turned the problem of excess biomass into an opportunity for the rural community. The effects extend far beyond the device. Less smoke in the air, less deforestation and a money making opportunity for rural farm areas producing fuel from waste.

This design was an Ashden award winner for 2002. Ashden.org for more information on their mission to elevate the world from the bottom up.

ARTI Director, AD Karve, reports producing activated charcoal by soaking sugar cane leaves in a 5% solution of calcium chloride and thoroughly drying before charring.

An•Thro•Soil•1 "Grassifier"

An•Thro•Soil•1 cooking supper while charring Indian Grass from old waste bales

Processing Complete

Developing a horizontal flow "trench" style smoke burning-char making system was a project at the Biomass Energy Foundation Camp in February 2011. The author had the pleasure of working alongside world champion stove builder Dr. Paul Anderson on the preliminary version of this design.

Although wood chips was the target fuel there, it turns out that horizontal flow is an excellent way of processing stemmy grasses into biochar.

Conceptually this is much the same as controlled burn of a field of grass into the wind. By containing the process in an oxygen starved environment, the downstream char is preserved instead of burned to ash.

The grass is lit under the chimney and the process moves toward the air intake on the opposite end. During processing, the smoke ignites right under the chimney creating a powerful draft and a noticeable roar. A short section of tube with openings cut in the center supplies the burst of oxygen that ignites the smoke.

When processing is complete and a flame plainly visible on the inlet end, the processor is completely sealed for cooling. Slide dampers shut off the inlet and chimney, caps seal the ends of the tube. The char is then allowed time to cool before opening the lid.

The inlet air is easily adjustable during runtime to control the rate of smoke production which sometimes overruns smoke burning capability. Inlet air control is also useful to control heat production on hot burning fuels like bamboo and cobs.

Processing and cool-down time varies greatly by feedstock:
- Wheat straw - typically 5 minutes processing, 5 minutes cool down
- Indian Grass (pictured) - 10-15 minutes processing, 10 minutes cool down
- Bamboo - 30 plus minutes processing, 15-20 minutes cool down
- Corn cobs - 40 plus minutes processing, 20 minutes cool down

Cool down doesn't mean cool to the touch, just cool enough to crush and remove the char wearing welding gloves. Opening sooner risks having some of the char catch fire.

The chimney would be at the bottom of this diagram

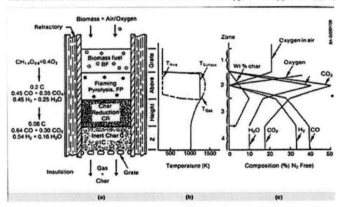

CHARRING VS BURNING

Provide just enough oxygen to thermally decompose and flare off hydrogen and CO leaving char.

The char cannot combust or reduce further in the absence of oxygen. No oxygen after pyrolysis.

Source - Biomass Gasification Handbook by Tom Reed and Agua Das available from Biomass Energy Foundation

Above is the classic stratified downdraft model for vertical processing of wood to char in insulated vessels.

Diagram is the same for fast pyrolysis of grassy feedstocks in a horizontal processor. Switch the above drawing from elevation to plan view without changing the drawing. No insulation and typically lower btu fuel limits internal process temps to the 500C range verified with thermocouple measurements. Secondary air below the chimney flares gases to burn the smoke and clean up emissions.

Design details and discussion

Feeding the smoke burning air was the biggest trick in this design. Virtually no pre-heat is available due to the short path of the air from outside to the chimney. The pattern below proved magical. (above)

Smoke burning tube air inlet configuration

Smoke burning air tube in action

Flame "race" can be a problem when trying to char grass horizontally. This problem was related by AD Karve of ARTI and noted by the author. "Race" describes fast paths through the fuel to the inlet air, that go so fast they skip processing portions of the feedstock.

Fuel packing and arrangement can have a large effect, particularly with corncobs. Ultimately a perforated plate at the inlet air end was a huge aid in preventing "race".

Overall this design is very useful for creating garden and food plot scale grass chars onsite.

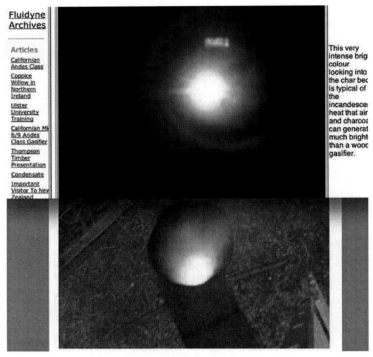

A look down the chimney of An•Thro•Soil•1 is eerily similar to that of an engineered charcoal gasifier

Processed cob versus original size

Processed Bamboo

MORE Discussion

The author has experienced fantastic results using grass chars for soils improvement. It may be due to grass chars being much easier to turn into fine powder, minerals in the grasses, or interior pore shapes different from wood. Whatever the reasons, grass chars seem to work well to improve soils, and are yet another form of available, often wasted biomass.

All chars are not created equal! The feedstock and processing make a difference. Research by Professor Kaneyuki Nakane of Hiroshima University discovered that bamboo char has almost five times the water absorption capacity of hardwood charcoal. Part of the magic of biochar in the soil is that it holds moisture bioavailable for plants. Bamboo char was presented by Professor Nakane as a preferred soil amendment for the challenging environment of rooftop gardening due to its high water holding capacity.

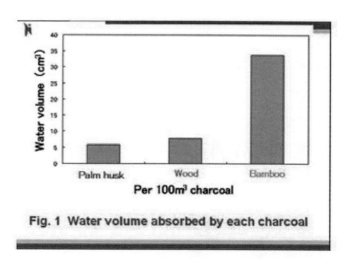

Fig. 1 Water volume absorbed by each charcoal

Source - *A New Proposition for Rooftop Gardening*, Kaneyuki Nakane

Fan Powered Smoke Burner

Chip Energy "Dragon" continuous feed heater and biochar maker

The **Chip Energy "Dragon"** eats biomass, streams an intense flame, and excretes biochar. It is the first smoke burner in this book to require an external source of energy besides lighting the fire. To light the fire, an accelerant soaked rag is lit and inserted through the lighting port (the chrome looking cap near the center of the picture).

Electrical power is used to power two augers and two small blowers. One auger feeds fuel, the other augers out biochar. One small fan controls the primary inlet air, the second controls the second shot of "smoke burning" air. Operating rates of all four components are independently adjustable by electronics to suit a range of fuel types and heat output. Maximum output is an honest 200,000 btu with turn down to 80,000 or lower while maintaining a smoke free process.

The other external power is a small propane flare used during startup or operational "issue" to assure the Dragon meets local codes of "no visible particulate emissions" exiting a structure.

The large metal box to the left of the upright pipe is a fuel feed hopper. Other than that, it is a simple system of a few smaller pipes crossing a larger vertical pipe.

The value proposition is free heat. Hardwood pellets convert to biochar at the rate of approximately four pounds of pellets to one pound of char. The biochar is sold for about four times the cost of the pellets by weight.

The Dragon eats roughly forty pounds of pellets per hour at maximum output, so a thousand pounds in the hopper provides unattended operation for a day.

The Dragon is yet another reminder that the simpler you make a smoke burner, the simpler it is to burn smoke. There is plenty of "free" energy available, for those who seek it.

The design scales to more output in increments of 200,000 btu. Chip Energy ships a 400,000 btu system with two Dragon's sharing a common hopper with all peripherals fully enclosed in a 20 foot shipping container.

An enclosed 200,000 btu Dragon system

June 2013 Chip Energy breaks ground on a project with dozens of 20 and 40 foot containers arranged into a processing facility to convert local low value biomass streams, from leaves to energy crops, into energy and useful processed products. The finished facility will feature one of the greatest carbon offsets of any such facility constructed to date. Anticipated opening in August 2013.

Chip Energy
395 W. Martin Dr. PO Box 85
Goodfield, IL 61742
309-965-2005 Fax 309-965-2905
Email pwever@chipenergy.com

Chapter 5
Burning Smoke in Internal Combustion Engines

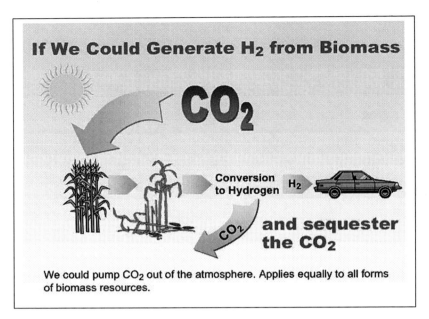

Source: Oak Ridge National Laboratories, Ag Wastes to Energy - 2003

There is no technological hurdle standing in the way of feeding a large round bale of switchgrass, wheat straw, or even corn stalks **directly** into a smoke maker to power a vehicle on the road. Just a lot of tinkering.

At an approximate rate of a mile per pound, there is about 2,000 miles worth of fuel on the bale bed pictured above. The bales took a few minutes to load. One bale is a thousand mile trip with room for a processor.

This may be a dream today. But a fuel just needs the right processor to make it work. Sawdust and wood chunks are already proven.

"Planting the second generation of biofuel feedstocks on abandoned and degraded cropland and low-input high-diversity native perennials on grassland with marginal productivity may fulfill 26—55% of the current world liquid fuel consumption, without affecting the use of land with regular productivity for conventional crops and without affecting the current pasture land." - **Land Availability for Biofuel Production, University of Illinois at Urbana—Champaign December 2010**

Although biomass can be converted into liquid fuels through a variety of processes, the additional transportation and processing to liquid typically consumes about forty percent of the energy during the conversion. Making smoke, burning smoke, on location is zero waste, maximum energy conversion, and maximum clean burning hydrogen.

Woodgas (smoke) powered engines run directly on vaporous gases, no carburetor required. Hydrogen and CO are the two largest flammable components of woodgas by volume. No fans or blowers are required after starting. In a symbiotic relationship, the suction created by each stroke of the engine pulls air through the woodgas system, making more smoke.

In 1927 Georges Imbert filed a patent, number 1,821,263 for what he called simply a **"Gas Producer"**.

"By the gas producer according to the present invention the problem of gas production from vegetable fuels for use with motor vehicles is effectively solved." ".. vegetable fuels such as wood, peat, stalks and the like, i.e. **out of fuel which is most widely distributed** and which under proper treatment yields gas and other power products of sufficiently high value to replace in a favorable manner the light hydrocarbons which are more or less rare and not easily recovered."

This of course was before "cheap oil" arrived on the scene.

Testing at Auburn University showed that at least one model of **modern** internal combustion engine is more efficient converting wood to power than converting gasoline. A Wayne Keith built system for a Dodge Dakota 318 cid traveled 168 miles on one million btu of gasoline, 231 miles on one million btu of wood.

Based on data in APPENDIX 1, Alabama Power Lab analyses (except for moisture content, which was not representative of these samples, and which was therefore determined separately at Auburn University) data presented in Table 3 was generated. It is concluded from these results that the gasifier was more efficient than gasoline. In particular, efficiency of wood was 37% higher than that of gasoline (231.6 miles per million Btu for wood, compared to 168.6 miles per million Btu for gasoline).

Table 3. Energy efficiency of different fuels based on distance traveled per unit of energy

Fuel	Total Btu Fuel	Total Mileage	Moisture %	Miles/gal gas equivalent	Miles/MMBtu
Gasoline	248,524	41.9	---	20.95	168.6
Wood	280,600	65.0	15.60	28.78	231.6
Wood + Broiler Litter	272,038	50.2	14.28	22.93	184.5
Wood + Switchgrass	298,050	58.0	14.56	24.18	194.6
Wood + Plastic	383,317	69.9	12.80	22.65	182.4

Source: **Feedstock Efficiency Tests with a small Pickup-mounted Down Draft Gasifier, July 2010**

Same as the stoves shown in earlier chapters, this is clean energy, with the primary exhaust products being carbon dioxide and water, no smoke. The fuel is short-cycle carbon, taken from the living stream and put back into the living stream.

A Note on Power Output

Even the best made conditioned smoke fuel has extraordinarily low energy density relative to liquid fuels like gasoline and diesel. Looking at the raw btu numbers, it is difficult to conceive how smoke can make acceptable power in an engine designed for refined petroleum.

The short explanation is charge density. The engine cylinder is filled with a relatively tiny shot of petrol per spark, a relatively much greater volume of woodgas per spark.

The quality of woodgas can be hugely variable, and is dependent upon the design of the processing system. The two designs described in this chapter make good "producer gas" that the author has witnessed producing quite useful power suitable for the original application in internal combustion engines originally designed for gasoline.

Low compression diesel engines converted to spark ignition may take best advantage of the unique characteristics of woodgas.

"The maximum de-rating in power is observed to be 16.7% in (wood)gas mode compared to diesel operations at comparable compression ratios. The extent of de-rating is much lower when compared to any of the previous studies (Parke et al., 1981; Ramachandra, 1993; Martin et al., 1981). This value matches with a similar kind of de-rating reported (Das &

Watson, 1997) with natural gas operation."
Facts about Producer Gas Engine, G. Sridhar and Ravindra Babu Yarasu December 2010 (Complete reference, including download link in the addendum)

MGS-80 Sawdust Gasifier

Missouri Gasification Systems S-80 Sawdust Gasifier

About 1980, Missouri Gasification Systems under contract with the Missouri Department of Natural Resources developed processors to convert sawdust and fine wood shavings into engine grade fuel. The S-80 was the smallest version, sized to power a small air cooled engine.

Making smoke, burning smoke is easy, once you know how to

do it. Turning flare quality smoke into engine grade fuel requires considerable downstream fuel conditioning.

To power an engine, the smoke must be cleaned and cooled. Fresh woodgas is a thick greenish tinted smoke that flares easily while hot. But it is full of moisture and potentially tars that could stick an engine valve.

The S-80 accomplishes the transformation from flammable smoke to useful fuel with cleanouts, baffles, and drains housed in a highly engineered rectangular box. Two larger sizes were manufactured, to power six cylinder and V-8 engines.

Operation

For the first start, the gasifier tube (under the lid on top left) is filled with crushed charcoal up the level of a ring of air intake ports. Sawdust is then poured in on top, up to the rim if desired, and the lid put back in place.

Next, the ignition port cap is removed, and a propane torch inserted to catch some char and sawdust on fire in the area of a ring of air supply ports.

Suction is applied at the gas outlet to get the operating temperature high enough to begin gasifying. A shop vac attached to the gas out port works well for fast starts. This sawdust gasifier is fast starting relative to most producer gas designs. It is typically just a few minutes from lighting to engine quality gas. The time from ignition to an engine quality flare was about 25 minutes with a small hand crank forge blower.

When the gas exiting the shop vac (with a flare attachment) starts burning with a clean flare (almost invisible in daylight), the shop vac hose is taken off, and the engine feed tube hooked up. Unless capped, smoke will pour out the intake air port while switching over. The engine feed tube is approximately the same diameter as the carburetor throat. A standard plumbing hose bib inserted before the the carb controls outdoor air mixing with the woodgas.

Modifying the engine timing, either with an after-market timing kit, or by cutting a new keyway about 15 degrees in the direction of shaft rotation, helps the engine start easily and run with good power on straight woodgas.

Ed Torrey of Torrey Enterprises tuning air/woodgas ratio

For extended operation, the sawdust fill cap is removed and more sawdust poured in. The pile is lightly tamped with a rod to break up any fuel bridging. This is done with the engine running. The operator learns how often to add sawdust. Refills and tamping were needed about every 30-45 minutes on this 13 horsepower engine powering a 6500 watt generator.

"in the field of alt energy, the sun does not always shine, nor the wind always blow, but smoke always rises" - **Jeff Schneider, moreenergy.org**

Design Notes

Incoming air is directed into a jacket surrounding the gasifier tube that the lid sits on. This preheats the intake air before injecting it through the ports.

After gasification, the gas proceeds downward through a grate, then up over a baffle into the first big cooling section above the "nasties" cleanout. The "nasties" are wet, tarry condensates that smell like a barbecue pit.

The next phase of conditioning is an oil bath with dipstick fill cap and drain. Vegetable oil or ATF has about the right viscosity. Then an easily removable final filter packed with wood chips or straw below, final filter media above.

A final baffle is traversed to aid cooling before the gas exits the system. Moisture is still condensing from the gas at this section, hence the drain. Amazingly, no external surface is dangerously hot to the quick casual touch.

During extended testing by Torrey Enterprises, the engine stuck a valve a couple of times. This didn't happen during running. It was when attempting to start the engine and evident by a noticeable lack of compression, spitting and backfiring.

Since this is a stationary setup, and space is not a concern, an additional large cooling section and much larger final filter tank filled with loose hay were added. With those additions, the valve sticking problem went the way of the dinosaur.

Why would anyone want to do this?
"Missouri Wood Gasifier" report, Missouri Department of Natural Resources, 1981

"According to figures from page 34 of a 1978 publication by the Missouri Department of Conservation, Forestry Section, entitled Wood Residues in Missouri; there are 847.22 million (mm) pounds of unused equivalent oven dry pounds of sawdust and shavings produced annually in Missouri. **This converts to 1,236,286 barrels of No. 2 fuel oil or about 1% of Missouri's average annual oil and gasoline consumption. This also converts to $37,088,580 at $30-per barrel.**

Over 90% of the money that Missourians spend on oil permanently leaves the state. For each $30,000 of energy dollars exported we lose one job to an oil producing state or country. This is 1,236 jobs lost because we are not utilizing these two waste products. This is a serious drain on Missouri's economy."

With a well designed and functioning gas producer, converting stationary power generation systems to burn smoke is more difficult for most people to comprehend than it is to actually do.

High quality conditioned "smoke" aka woodgas, is similar in btu, moisture content, and other physical characteristics to natural gas. Natural gas engines are available factory built.

Propane engines are easily adapted to woodgas, using the same techniques described above for the small air cooled engine. First figure out a way to supply vaporous fuel into the engine. Then with gas hooked up, adjust ignition timing until the engine runs smoothly with good power.

Torrey Enterprises fabricated a gas producer similar to the Wayne Keith system described next, to power a formerly propane generator. The engine was a small 4 cylinder Ford turning a generator that easily supplies a large house with all modern electrical conveniences.

The S-80 is a tidy, well-engineered package. But with stationary systems, space is not usually an issue. In the propane conversion, large, tightly sealed containers and pipes served the same functions of cooling and cleaning as the tightly packaged S-80. A heat exchanger section was added to pre-heat incoming air, pre-cool outgoing gas.

After an initial hiccup caused by a moisture entrainment problem, the former propane system ran continuously almost eight hours until boredom got the best of the testers. Feeding wood was the only action required during the run. Draining condensate and dumping char would be required for extended operation with the frequency determined by the size of the containments for those items.

Stationary biomass power systems are in much broader use than most people realize. In remote areas with plentiful biomass resources and high priced diesel, they offer the fastest payback in alt energy.

The Phillippines is one such area where the technology is taking hold.

"We see great applications all over the country. As fossil fuels like diesel increase in price, we see biomass gasification as the best alternative. We can cut the energy costs of existing diesel users by 50% **with a payback of less than 1 year**," according to Engr. Allan Reyes of Magco. The Dinalungan power plant is a live demonstration of how biomass power can help bring down energy costs and provide power in remote areas. "

- First Biomass Gasifier for Rural Electrification in Philippines inaugurated by Gov. Angara-Castillo - **PRLog.org March 2012**

Smaller power plants and de-centralization lessen the catastrophe potential of losing one big power plant to typhoons or earthquakes. Smaller plants are also easier to bring back online following natural disaster.

The Wayne Keith Gasifier for Carbureted and Fuel Injected Engines

Powering vehicles is the greatest challenge in smoke burning
- Fuel and processor must fit onboard
- Must be able to operate over a wide range of engine speeds
- Operator training
- Must retrofit to an engine not originally designed for woodgas

Doug Brethower, Jeff Schneider and woodgas powered 460cid Ford pickup

In the world of woodgas powered vehicles, there are many rabbit trails, few rabbits. Sir Wayne Keith of Springville, Alabama is THE rabbit of woodgas powered modern vehicles. "In theory there is no difference between practice and theory, in practice there is."

Wayne is the world leader in miles driven, world land speed record holder, and has a single continuous trip of almost 8,000 miles on woodgas in his list of credits.

Sir Wayne drew a line in the sand at $1.50 per gallon, and said "from now on the buck stops here, in my pocket." His travels since about 2005, across town, and across the North American continent have primarily been powered by waste wood from his sawmill.

The imbert was a great gas producer for engines of that day. Wayne's design improves upon the original concept with a design that works equally well on modern fuel injected engines. The red Ford pictured above was his original test bed beginning in 2004. It has been replaced for farm chores with a 2004 V-10 Dodge. For cross country running, a Dodge Dakota with 318 cid engine is his platform of choice. Both are fuel injected, actually dual-fuel capable of running on gasoline or woodgas with switch over accomplished from the driver position in the cab.

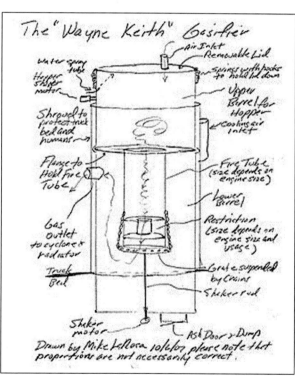

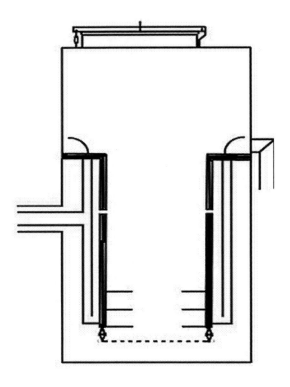

Wayne Keith Gasifier Drawings by Mike Larosa 2007 (previous page) and Jim Hart 2011

The smoke making section of the Wayne Keith system is a breakthrough. It is not an imbert, rather characterized as a stratified downdraft. Wood is filled at the top. The heat of the process below essentially conditions the fresh dry wood into nearly charcoal by the time it reaches the inlet ports.

The curled lines in the Jim Hart drawing indicate a "gutter" that catches moisture and condensate running down the uninsulated walls of the upper hopper. This drains into a catch tank through the pipe exiting right. Process heat and hot gas surround the incoming air to the greatest extent possible. Fresh process air travels through an innovative system of pipes and baffles that is surrounded by exiting hot gases on the other side of the walls. This helps maintain metal parts temperatures below metal melting temperatures. No exotic materials are used or required.

This innovative baffling arrangement also maximizes intake air pre-heat while cooling the hot exiting gases, both desirable. The vertical center of the gasifier is where wood turns to char then to hot gas. Tracing the route of intake air from the gasifier ports (the holes in the innermost "firetube") back to the interior pipe on the left shows how thoroughly this design principle is followed.

After exiting the gasifier proper, another round of air to gas heat exchange takes place in a dedicated heat exchanger section.

By the time the gas enters the condensing rails that look like custom pickup bed rails, the gas has been cooled to about 400 degrees F. The bed rails are designed to cool the gas to outdoor air temperature by the time the gas exits the rails at the back of the pickup bed.

From that point standard PVC plumbing connections are used to move gas into a final filter section before fuel grade vapors are pulled into the engine by it's own suction. A large condensate catch tank is fastened below the back bumper to catch the considerable moisture wrung from the gas as it cools in the bed rails.

By the time the gas hits the condensing rails that look like custom bed rails, the gas has been cooled to about 400 degrees

F. From that point on, it is more or less standard gas cooling and filtering to finish conditioning the vapors into engine grade fuel.

There are of course a lot more details involved. The details right down to the finer points of barrel welding are covered in a book *"Have Wood, Will Travel"* by **Wayne Keith and Chris Saenz**. A far cry from previous works on woodgas for engine fuel, HWWT focuses on simple, practical instructions for fabricating a rigorously proven design. Even if you plan on using a different gasifier design, the tips in the book for sizing gas flow to engine demand and driving a woodgasser are priceless. The book is available at driveonwood.com

Wayne Keith woodgas powered Dodge Dakota in Pittsburg, Kansas

Salvage barrels are Wayne's favorite shell material. Metal barn siding filled with ashes is a favorite insulation. Used ammo cans work well for tightly sealed, easily opened "ash dumps". It is not actually ashes, rather a high grade of charcoal that comes out the ash dumps. Add worlds fastest charcoal kiln to Sir Wayne's list of credits. The simpler you make it, the simpler it is.

Wayne Keith Gasifier base - a barrel, barn siding, and ash dumps welded in the bottom for periodic charcoal cleanout

Mike LaRosa - Linden, Wisconsin

Wayne's good friend Mike LaRosa has been traveling on woodgas periodically since the 1970's and the "tumbleweed experience". Mike's designs have a decidedly more refugee appearance. Mike is an expert at making "good gas" from parts put together in a hurry. His favored air intake port system, notorious in woodgasser circles as "the LaRosifier", is made from a salvaged vented brake rotor. From there he builds the hopper up, and the gasifier down, based upon what he has laying around or will function with a minimum of fabrication. Wayne estimates about 250 hours of labor to build his art. In a few days Mike can be making good gas from "parts".

A LaRosa gasifier system before fitting condensing rails and gas out to engine

"The extraction and burning of petroleum is an abomination, the purpose of which is to enslave humanity to a pump and hose in lieu of a ball and chain" - iRobert

Afterword

"The technological advancements that make contemporary society possible are the result of some ten thousand years of development of the intentional use of fire. Yet there is surprisingly little information on the practice and importance of pyrotechnology" - **Mastery and Uses of Fire In Antiquity, J. E. Rehder**

The world needs energy. Every day, the sun provides more energy than humanity has even begun to learn how to use. Marching on Washington will not bring the quadrillions of btu of safe, clean, environmentally durable biomass energy currently unused to market.

Real change requires people really changing. The knee jerk reaction of the masses is to deny and dismiss disruptive forces as long as they possibly can. Real change almost by definition starts with a few fringe lunatics that make magic happen. The masses ask "why would anyone want to do that"? But a few followers join in. Then a few more. This book is written in hopes it will touch the "few more".

When the "few more" start to join in is when real change happens. That is when some in the masses start fearing they may be missing out, and start thinking in terms of what must be possible, because it is being done. Then change can happen in a hurry. Soon that change is accepted as the way things have always been.

The short history of touch screen phones gives some idea of how fast change can happen when real change is mass deployable.

"Biomass represents an important but underutilized energy resource in the United States. The Congressional Office of Technology Assessment has estimated that, with proper resource management and the development of efficient conversion processes, the potential contribution of biomass to U.S. energy demand could range as high as 17 quadrillion Btu per year--almost 20% of current U.S. energy consumption." -

Biomass Thermochemical Conversion Program, 1983 Annual Report - Pacific Northwest Labs DOE Battelle

Thirty years later, the price of oil has tripled, the biomass remains plentiful, and technologies for conversion have advanced dramatically. The only thing holding it back today is perception and waiting for somebody else to do it. Seventeen quads of energy is equivalent to almost 3 billion barrels of oil.

Addendum

A. US Agricultural Wastes

Source: *Agriculture Waste to Energy* - **Robin L Graham, PhD**
Oak Ridge National Laboratory, June 2003

A "quad" (quadrillion btu) = 172 million barrels of oil

B. Biomass Stoves Emissions Testing

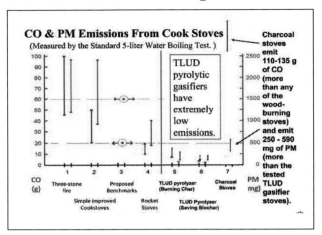

Source: *Bioenergy Issues of Cookstoves for Challenged Societies and Environmental Improvement*
Paul S. Anderson, PhD, Hugh S. McLaughlin, PhD, PE
October 2010

The proposed benchmarks from PCIA, Partnership for Clean Indoor Air

---

C. Biochar and Durability of Carbon in Soil

"The half-life of C in soil charcoal is in excess of 1000 yr. Hence, soil-applied charcoal will make both a lasting contribution to soil quality and C in the charcoal will be removed from the atmosphere and sequestered for millennia."
David Laird, National Soil Tilth Laboratory, Ames, Iowa
The Charcoal Vision
Agronomy Journal • Volume 100, Issue 1 • 2008

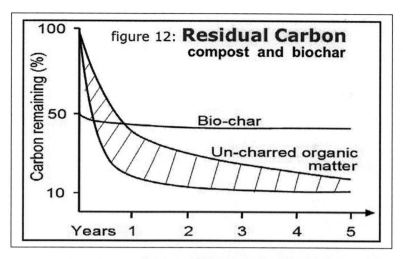

Source: *Carbon-negative Farming* - David Yarrow, December 2010

"The oldest description on charcoal use in agriculture is found in a text book, "Nogyo Zensho (Encyclopedia of Agriculture)" written by Yasusada Miyazaki in 1697 (7). He described in it; "After roasting every wastes, the dense excretions should be mixed with it and stocked for a while. This manure is efficient for the yields of any crops. It is called the ash manure ". Probably similar knowledge had been popular in China and Korea since ancient time."

Source: Introduction to the Pioneer Works of Charcoal Uses in Agriculture, Forestry and Others in Japan
Dr. Makoto Ogawa September 2010

D. FEMA GASIFIER 1989

Credited with raising awareness, but not known as a particularly good gasifier design.

D. CLEW Diagram Camp Lejeune Energy from Wood biomass plant.

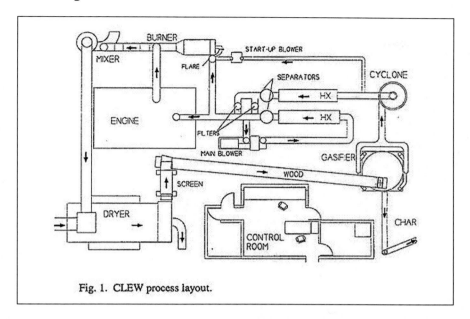

Fig. 1. CLEW process layout.

"The CLEW plant utilizes only wood waste from the Base, which by regulations has to be diverted from the landfill. Utilizing waste biomass as a fuel for power generation eliminates sulfur dioxide emissions; produces zero net gain of carbon dioxide (CO_2); minimizes waste disposal problems, tipping fees, and the purchase of fossil fuels for electricity; provides energy security at domestic and international military installations; and promotes an exportable technology."

Source: DEMONSTRATION OF A 1 MWe BIOMASS POWER PLANT AT USMC BASE CAMP LEJEUNE, J. Cleland & C.R. Purvis August 1997

1 MWe is widely reported as roughly sufficient energy to power 1000 "average" houses.

E: Interconnected Power Grid Vulnerable to Severe Space Weather

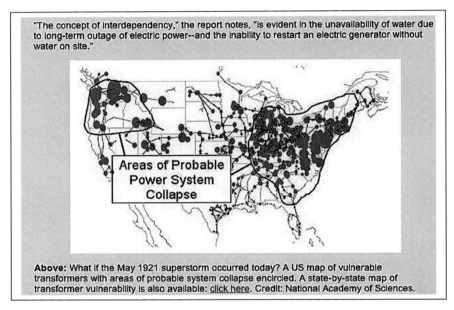

"The concept of interdependency," the report notes, "is evident in the unavailability of water due to long-term outage of electric power--and the inability to restart an electric generator without water on site."

Above: What if the May 1921 superstorm occurred today? A US map of vulnerable transformers with areas of probable system collapse encircled. A state-by-state map of transformer vulnerability is also available: click here. Credit: National Academy of Sciences.

Source: Severe Space Weather--Social and Economic Impacts, NASA Science News, January 2009

"Did you know a solar flare can make your toilet stop working?"

F: Global Atmospheric CO2 readings 1958-June 2013

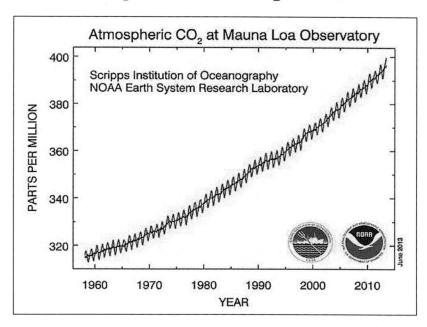

**Trends in Atmospheric Carbon Dioxide
National Oceanic and Atmospheric Association
http://www.esrl.noaa.gov/gmd/ccgg/trends/**

**Recommended further reading on the subject of stoves:
Understanding Stoves**

For Environment and Humanity 1st Edition, released 2012

(OK) This is an 'Open Knowledge' book as declared by the author.

Dr. N. Sai Bhaskar Reddy, 2012 saibhaskarnakka@gmail.com
  http://goodstove.com

Micro Gasification: Cooking with gas from biomass

1st edition, released January 2011

Author: Christa Roth

Published by GIZ HERA – Poverty-oriented Basic Energy Service

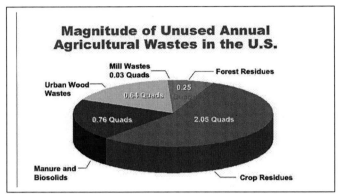

On the WWW (World Wide Web)

Aprovecho Research Center - Improved Stoves for the Developing World
 -- http://aprovecho.org
Producer Gas for Engines

Facts About Producer Gas Engine, Paths to Sustainable Energy, Dr Artie Ng (Ed.)

G. Sridhar and Ravindra Babu Yarasu (2010)

ISBN: 978-953-307-401-6, InTech, DOI: 10.5772/13030.

Available from: http://www.intechopen.com/books/paths-to-sustainable-energy/facts-about-producer-gas-engine

The Final Word

by David Yarrow

LIFE: **L**ocally **I**ntegrated **F**ood and **E**nergy

"Food and energy are the two keystones of any community anywhere on earth.

If we produce and distribute food and energy locally, we have:
- the food
- the energy
- and the money

We establish the capacity to create and retain wealth in our community.

We put in place the two foundations of any human economy."